ISBN 978-1-332-27073-6
PIBN 10307110

English
Français
Deutsche
Italiano
Español
Português

www.forgottenbooks.com

Mythology Photography **Fiction**
Fishing Christianity **Art** Cooking
Essays Buddhism Freemasonry
Medicine **Biology** Music **Ancient**
Egypt Evolution Carpentry Physics
Dance Geology **Mathematics** Fitness
Shakespeare **Folklore** Yoga Marketing
Confidence Immortality Biographies
Poetry **Psychology** Witchcraft
Electronics Chemistry History **Law**
Accounting **Philosophy** Anthropology
Alchemy Drama Quantum Mechanics
Atheism Sexual Health **Ancient History**
Entrepreneurship Languages Sport
Paleontology Needlework Islam
Metaphysics Investment Archaeology
Parenting Statistics Criminology
Motivational

WORKING PAPER
ALFRED P. SLOAN SCHOOL OF MANAGEMENT

THE MOTIVATIONAL IMPACT OF MANAGEMENT-BY-EXCEPTION

IN A BUDGETARY CONTEXT

by

Peter Brownell

MASSACHUSETTS

THE MOTIVATIONAL IMPACT OF MANAGEMENT-BY-EXCEPTION

IN A BUDGETARY CONTEXT

by

Peter Brownell

WP 1311-82

May 1982

Revised October 1982

1.0 Introduction

It is often suggested in the managerial accounting literature that control system effectiveness depends not only on what information the system supplies to managers responsible for exercising control, but how managers use the information. In the context of budgetary controls, Argyris (1952) may have been the first to note that control system failure was more often affected by the style of information use rather than by the technical characteristics of the system. Subsequently, several empirical studies have focused attention on this issue (e.g., DeCoster and Fertakis [1968]; Hopwood [1972]; Otley [1978]). At the practical level, one common style of information use is "management-by-exception" (MBE).

Exception-reporting is probably based on the notion that only variances, both unfavorable and favorable, should attract managerial attention, particularly if they were assessed as being significant. Hence, MBE provides a basis for allocating, or rationing, managerial effort.

In practice, however, such systems typically reduce to a biased preoccupation with unfavorable variances. For example, Bittel (1964, p. 5) defined MBE as system of control which enables the manager to be "spared the task of reviewing performance where things are going well (so as to) devote his attention only to those areas which really require his managerial attention." Ronen and Livingstone (1975, p. 680) suggested that, under MBE, "the response to favorable deviations not requiring corrective actions often seems to be weaker than that to unfavorable deviations." Birnberg and Nath (1967, p. 478) described exception reporting as a system in which "the emphasis in the feedback is on punishment rather than some mixture of punishment and reward," while Porter and Zannetos (1978, pp. 12-13) characterized the phenomenon as one by which managers are given "full blame for failures and less than full credit for successes." Itami (1975) provided an analytical demonstration of how management, in the design of subordinate reward structures, might purposefully employ asymmetrical rates of penalty and reward (for unfavorable and favorable budget variances, respectively) so as to "modify" the subordinates' attitudes towards risk and thereby increase the likelihood of goal-congruent behavior on the part of the latter. He goes on to suggest that "this is perhaps common practice in many budgetary control systems."(p. 85) Baiman and Demski (1980) obtained similar results and, like Itami, concluded that "Certainly the low-tail conditional variance investigation system is often observed in practice (It is probably the system which is most often used)."(p. 194, parentheses in original)

Except for these last results, the general assumption in the other studies is that this variant of MBE has adverse behavioral consequences. However, none offers any systematic empirical evidence of such effects. The purpose of this paper is to report the results of an empirical study of the effects of MBE in a budgetary context on individual motivation, where MBE is assumed to refer to a preoccupation with unfavorable variances. I initially hypothesize that MBE will have adverse effects on motivation (the general hypothesis), but that these effects will be mitigated by the level of participation in budget setting (the contingency hypothesis).

As suggested by the variety of literature discussed above, a theoretical framework based on the behavioral sciences, as well as one based on the economics of agency, were possible candidates for the present study. A behavioral framework was chosen, but later in the paper I attempt to link the framework (and the more important results of the study) to some of the propositions which have emerged from the economic perspective.

In the next section of the paper, I develop the theoretical framework to support the general hypothesis. The section following elaborates on the role of budgetary participation in moderating the effects of MBE (the contingency hypothesis). Sections dealing with the method, results, and conclusions then follow in turn. One interesting finding is the strong evidence in support of the above characterization of MBE as being preoccupied with unfavorable budget variances. The results also provide some support for the contingency hypothesis, but not for the general hypothesis.

2.0 A Theoretical Framework - The Expectancy Model of Human Motivation

The literature in psychology dealing with the acquisition, maintenance and modification of behavior falls into two broad classes. First, there is the behaviorist perspective, characterstic of early learning theorists (e.g., Hull [1929]; Spence [1942]), in which learning and motivation are conceived solely in terms of the linking of stimuli and responses. The theories of classical and operant conditioning are two well-known conceptualizations of learning and motivation which have derived from a behaviorist tradition.

A second, more recent perspective can be referred to as the cognitivist view. Here, the intervening cognitions (between stimulus and response) are the focus of attention. While this perspective can be traced back at least as far as Tolman [1932], the more recent work of Rotter [1954], [1966] and Bandura [1962] [1965] [1972] has integrated the roles of beliefs, expectations, and other cognitive processes into extensions of traditional learning theory. Rotter [1966] suggests that the occurrence of a particular behavior is dependent not only on an individual's objective reinforcement history (the behaviorist view), but also on the individual's expectation that the behavior will result in a particular consequence and the subjective assessment of the costs and benefits associated with the consequence (the cognitivist view). These constructs are readily recognizable in the expectancy theory conception of motivation, which provides the theoretical linkage between MBE and motivation underlying the choice of the criterion variable used in this study.

MBE, with its stress on unfavorable variances, can shape both the objective reinforcement history of the individual, and the subjective expectancies associated with any new behavior. In either case, motivation will suffer. This can be seen by examining the particular formulation of the expectancy value model of motivation chosen for this study.

The formulation, based on theoretical work of Georgopoulos, et al (1957) and subjected to empirical test by Galbraith and Cummings (1967), can be expressed as follows:

$$M = IV_b + P_1(IV_a + \Sigma P_{2i} \ EV_i) \qquad (1)$$

where

M = motivation to work.

IV_a intrinsic valence associated with successful performance of the task (goal-accomplishment).

IV_b intrinsic valence associated with goal-directed behavior.

EV_i = extrinsic valence associated with i^{th} extrinsic reward contingent on work-goal accomplishment.

P_1 the expectancy that goal-directed behavior will accomplish the work-goal (a given level of specified performance).

P_{2i} the expectancy that work-goal accomplishment will lead to the ith extrinsic reward.

MBE is hypothesized to adversely affect the P_{2i} values in the above formulation, since those P_2's associated with positively valent outcomes,

4

or reinforcers, are expected to be assessed systematically lower under MBE due to an absence of historical association between goal-accomplishment and the receipt of positive reinforcers.

For similar reasons, IV_a might also suffer a deterioration under MBE, although a dissonance-resolving process, such as that suggested by Calder and Staw [1975], may mitigate this effect. That is, individuals who cannot attribute their goal-directed effort to the expectation of extrinsic rewards because few such rewards are forthcoming under MBE, might rationalize their expenditure of effort in terms of the internalized value of goal-accomplishment, or IV_a. Therefore, although overall motivation may not suffer, the role of intrinsic versus extrinsic rewards in contributing to overall motivation will shift. The use of the expectancy theory model of motivation chosen for this study permits an examination of the association between MBE and specific elements of the model.

Finally, reduced assessments of P_1 may occur because individuals may perceive the non-receipt of extrinsic rewards as resulting also from a failure of the reinforcing agent to acknowledge goal-accomplishment itself. Such a phenomenon translates into reduced assessments of P_1.

But, here again, reduced assessments of P_1 may be offset by increased assessments of IV_a. The intrinsic value of goal-accomplishment can be viewed as being a positive function of goal difficulty, and, hence, a negative function of P_1. Indeed, early conceptions of motivation viewed IV_a as a simple negative function of the probability of success (e.g., Atkinson [1958]).

To summarize, MBE is hypothesized to impact motivation through reduced assessments of the P_{2i} and P_1, with the possibility, in both cases, of compensating adjustments to IV_a.

3.0 Budgetary Participation - The Contingency Hypothesis

The effects of budgetary participation have been widely examined over the years by researchers in managerial accounting, although not with consistent results. The findings range from strong positive effects of participation on performance and job satisfaction (e.g., Kenis, [1979]), to weak positive (e.g., Milani [1975]), and even to negative associations (e.g., Bryan and Locke [1967]). The conclusion is that the effects of participation are contingent on a variety of moderating effects (Brownell [1981], [1982]). Indeed, the results of some studies provide a basis for predicting interactive effects of participation and MBE.

Cherrington and Cherrington [1973], in a laboratory experiment, examined the interaction between budgetary participation and reward structure. The latter was reflected in the degree of reliance placed on budgeted results in performance evaluation. Their evidence was mildly in support of the hypothesis that both performance and job satisfaction would be higher when a high reliance on budget information was accompanied by high participation. One reason for this is that in providing the individual with the opportunity to influence budgeted results, management achieves more legitimate grounds for strong sanctions in the event of sub-standard budget-related performance.

In a similar type of investigation, Brownell [1982] examined the role of budget participation as a moderator of the effects of different superior evaluative styles. Two prior investigations of the impact of evaluative style had produced contradictory results (Hopwood [1972] and Otley [1978]). In his study, Brownell found results which supported the hypothesis that a heavy reliance on accounting information, and on budgets in particular, would not adversely affect performance as long as this evaluative style was accompanied by a high level of participation in budget setting.

6

The common thread in both of the above studies is the notion that participation in budget setting may alleviate otherwise adverse consequences stemming from a style of use of accounting information which places a high emphasis on budget achievement in performance evaluation and reward administration. A similar effect of participation is hypothesized in connection with the use of MBE.

4.0 Hypotheses and Method of Study

4.1 Hypotheses

The above can be summarized in the two hypotheses to be tested in this study. In null form we have:

H_1 There will be no relationship between MBE and motivation.

H_2 There will be no interaction between MBE and budgetary

participation affecting motivation.

4.2 Method

Data for the study were collected through a survey questionnaire to 224 middle-level managers drawn from three separate corporations – two in the electronics industry and one in the steel industry. The managers were drawn from a variety of functional fields including marketing, production, research, and administration. The one selection criterion was that the activities of the managers should be controlled via the use of budgeting. However, final sample selection was left up to top management in each corporation and so the sample was not strictly random.

Of the 224 questionnaires distributed, 140 were returned (62.5%), of which 122 were usable. The average age of respondents was 37.0 years and their average tenure with their respective companies was 8.1 years.

Measures of three variables were obtained in the questionnaire: motivation, MBE, and budget participation.

4.2.1. Motivation

The objective of the questionnaire items was to elicit measures on each construct of the model presented in Equation 1. For expositional purposes, the constructs of Equation 1 are classified into two basic groups valences, and instrumentalities/expectancies.

Valences: The approach to the measurement of the three classes of valences (IV_b, IV_a, and EV_i) was adapted from the procedure developed by Lawler and Suttle (LS) (1973) in such a way to distinguish between the three classes of valence. Seventeen outcomes from the LS set were used in the present study and these were a priori classified as either intrinsic (8) or extrinsic (9). The outcomes are listed in the appendix. For each outcome, respondents were twice asked to indicate on a scale from one to nine (extremely desirable to extremely undesirable)[1] the strength of their preference for that outcome. First, respondents were asked to value each outcome as it might result from "working hard" (goal-directed behavior), and second, to value the outcomes as they might result from "meeting or beating budgeted goals" (goal-accomplishment). IV_b was measured by averaging the eight responses to the intrinsic items from the first set of responses. IV_a was measured by averaging the responses to the same eight items, but from the second set of responses.

Two points should be noted here. First, IV_b and IV_a are incorporated into the expectancy model by way of a single average score, as implied by the model presented in Equation (1). Nonetheless, it is difficult to ascertain why theoretical discussions of the expectancy model have overlooked the issue of multiple intrinsic valences, and hence the possibility of their entering the formulation by way of a summation. The two options (mean versus summation) are clearly different in that they place

8

different weights on the intrinsic versus extrinsic elements of the model.

Second, the use of the same eight intrinsic outcomes in the measurement of IV_b and IV_a presupposes that any given outcome could result from either goal-directed behavior or goal-accomplishment and that the value placed on the outcome (valence) might not be independent of its source.

Table 1 presents the means and standard deviations of the eight outcomes (after scale adjustment) for each of IV_b and IV_a. The degree of independence of the two sets is reflected in the coefficients of correlation among the two sets (last column) and the t-statistics for eight paired difference of mean tests. The results reveal that while a significant degree

INSERT TABLE 1 HERE

of association exists between the two classes of intrinsic valence, their respective mean scores were usually quite different.

Turning to a consideration of the EV_i, the valences associated with the nine extrinsic outcomes were assessed from the second set of responses (outcomes resulting from goal-accomplishment). Table 2 presents the means and standard deviations of the nine extrinsic valences.

INSERT TABLE 2 HERE

9

Instrumentalities/Expectancies: Equation (1) calls for the assessment of nine specific instrumentalities associating goal-accomplishment with extrinsic outcome (P_2). Again following LS, these instrumentalities were assessed by asking respondents to indicate on a scale from one (never) to seven (always), how often "meeting or beating the budget" would result in each outcome. Three additional items incorporated in this set of questions elicited measures of P_1, the probability or expectancy that goal-directed behavior would result in goal-accomplishment. For ease of interpretation, the twelve responses were converted to probabilities in the range of zero to one and the means and standard deviations of the nine P_2's are presented in Table 3. The three assessments of P_1 were significantly $(P < 0.01)$ correlated and the mean correlation was $+0.37$. A single measure of P_1 was therefore derived by averaging the three responses. The three correlations, and the final mean and standard deviation for P_1 are also presented in Table 3.

INSERT TABLE 3 HERE

The final measure of motivation was obtained by aggregating the individual constructs following Equation 1. Table 4 reports descriptive statistics on the final measure.

INSERT TABLE 4 HERE

10

4.2.2. Management-By-Exception (MBE)

A review of previous literature uncovered no suitable empirical measure for MBE. The measure developed for this study involved nine questionnaire items. One item simply asked respondents to rate on a seven-point scale the frequency (one equals never, seven always) with which unfavorable variances receive more attention than favorable variances. The remaining eight items consisted of four pairs of similar items. Four items relating to the investigation of variances were taken from the so-called Budget-Related Behavior questionnaire (see, for example, DeCoster and Fertakis, 1968; Bruns and Waterhouse, 1975; Swieringa and Moncur, 1975; and Merchant, 1981). An example is "My superiors discuss budget items with me when variances occur." Each of four such items were included twice in the questionnaire, once for favorable variances, and once for unfavorable variances.

Since the definition of MBE used here involves managerial assessments which are pre-occupied with unfavorable variances, the response to the "favorable variance" version of each item was subtracted from the response to the "unfavorable variance" version of the same item, to obtain four difference scores. All scores were converted to probabilities, and Table 5 reports first the mean scores for the single, overall assessment, followed by the means for each of the four pairs of specific items, their differences, and the significance of those differences (paired test).

INSERT TABLE 5 HERE

As noted in the introduction, the results presented in Table 5 wholly

11

support the empirical observation that MBE, as defined in this study, is quite prevalent in practice. All four of the pairwise comparisons among the eight specific questionnaire items reveal a significantly higher frequency of the "budget-related behavior" as it relates to unfavorable variances than as the behavior relates to favorable variances. Roughly interpreted, the result for the single item addressing MBE, indicates that, on average, more attention is given to unfavorable variances than to favorable variances 81% of the time. Given the consistent pattern of results from the entire nine item instrument, I constructed a single average measure of MBE using the response to all nine. To do so, I summed the differences between the four paired items plus the single item score,[2] and then divided by five.[3]

These results are consistent with a suggestion of Baiman and Demski (1980) who provided a convincing rationale for the prediction that MBE, as defined here, is prevalent. As previously mentioned, they derived an analytical result which suggests that, for monitoring risk-averse managers (agents), top management (principals) would do well to employ an investigation strategy which focuses on unfavorable variances. Beyond this, they offer (P. 194) an appealing argument suggesting the greater likelihood of encountering risk-averse individuals among middle-level management (the objects of monitoring – and the subjects of this research) than among top-management (the monitors themselves).

4.2.3. Budgetary Participation

The participation measures developed by Milani (1975) and Hofstede (1967) were both employed in this study, as a crude means of validating one another. Previous use of both measures in a single study (Brownell, 1982) revealed that the two measures correlated significantly (0.74).

The Milani measure is a six item Likert type scale, each item calling

12

for a response from one to seven. The scale is designed for an additive construction of the overall score, and a previously performed factor analysis of the scale (Brownell, 1982) provides adequate confirmation of the single factor nature of the measure.

The Hofstede measure is an eight-point, fully anchored single scale calling for one response. In my previous use of both measures, I relied on Milani's for hypothesis tests because of the validation of its structure. But since the two measures correlated at only $r = 0.59$ in this study, hypothesis tests concerning budgetary participation were performed using both measures. Table 6 reports descriptive statistics for each.

INSERT TABLE 6 HERE

5.0 Results

H_1: The first hypothesis proposed an overall negative relationship between MBE and motivation. The simple correlation between the measures yielded a coefficient $r = -0.09$ which, while in the predicted direction, is not statistically significant. Hence, the null hypothesis of no relationship cannot be rejected.

In order to further explore this result, MBE was correlated separately with each of the eight intrinsic valences (both IV_a and IV_b), the nine P_2's, the nine EVs and the three components used to construct P_1. The results are presented in Table 7.

13

INSERT TABLE 7 HERE

Taken individually, the coefficients in Table 7 fail to uncover any compensating effects of MBE on the separate components of the model. Taken jointly, however, the results do offer some support for the theoretical propositions offered earlier. In particular, of nine correlations between MBE and each of the nine P_2's, eight are negative (one significantly), as expected (see page 4). A binomial test indicated that the probability of this occurrence is less than 0.02. Note also that while no particular relationship existed between MBE and either IV_a or EV_i (both contingent on goal-accomplishment), some evidence of a positive relationship between MBE and IV_b was found. Six of the eight correlations are positively signed ($P < 0.11$) and one is highly significant. This result provides tentative support for a slight variation of the dissonance-resolving process discussed earlier. There I suggested that MBE could operate to shift valued outcomes from those contingent on external agents (EV_i) to those internal to the individual (IV_a and IV_b). It seems, however, that the outcomes associated with goal-directed behavior (IV_b) as opposed to goal-accomplishment (IV_a and EV_i), are more highly valued under MBE. Finally, all three assessments of P_1 are negatively associated with MBE, implying that the use of MBE is associated with increased subjective assessments of goal-difficulty. This possibility was also alluded to in the earlier discussion.

H_2: The second hypothesis was that increased budgetary participation would favorably influence the relationship between MBE and motivation. As a first test of this hypothesis, the two sets of participation scores were dichotomized into their upper and lower halfs, and the correlation measures computed between MBE and motivation observed under each participation condition (high versus low). Table 8 reports the results.[4] The correlations under high participation are approximately zero, while under low participation they are both significantly negative ($P < 0.05$, one-tail test). The results are consistent for both measures of participation. Using the Fisher transformation (see, for example, Winkler and Hays [1975] p. 653), the two correlations under each participation measure differ at $p < 0.05$, providing support for the rejection of H_2.

INSERT TABLE 8 HERE

A more efficient means of testing H_2 involves a regression of motivation on MBE, the dichotomized participation scores ($+1/-1$), and a multiplicative interaction term. The coefficients from this regression can then be used to estimate the intercept and slope, respectively, of two simple regressions of motivation on MBE - one for each of the two participation groups.

The regression equation was:

$$Y = \alpha + \beta_1 X_1 + \beta_2 X_2 + \beta_3 X_1 X_2, \tag{2}$$

where Y = Motivation

X_1 = MBE, rescaled, mean = zero.

X_2 = Budgetary Participation (+1 = high, -1 = low)

In this model, high values of MBE (positive) will combine with +1 for

15

high participation to produce a positive interaction term. Low values of MBE (negative) will combine with −1 for low participation, also to produce a positive interaction term. Motivation should generally be higher in these conditions, so a positive interaction coefficient, β_3, was expected.

The results of the regression are presented in Tables 9 (for the Milani participation measure) and 10 (for the Hofstede measure).[5] Both analyses reveal positive interaction coefficients, but each is significant at only the 0.10 level. These results are consistent with the above correlation tests.

INSERT TABLES 9 AND 10 HERE

As a final analysis, the coefficients from the regression using the Milani measure (Table 9) were utilized to reconstruct a functional relationship between motivation and MBE separately under high and low participation:-

High participation: $Y = 10.46 + 0.35X_1$ (3)

Low participation: $Y = 9.46 - 6.11X_1$ (4)

These two functions intersect at $X_1 = -0.15$. Since X_1 has a mean of zero in Equation 2, this point of intersection is $X - 1.24\sigma_X$, indicating that it is somewhat unlikely (at least on the basis of the present sample) to

16

encounter a score on MBE which is sufficiently low for motivation to be higher under low participation. Indeed, the significant, positive coefficient β_2 (on participation) indicates that, in general, higher participation is associated with higher motivation.

6.0 Conclusions

The results of this study were generally disappointing. While the overall pattern tends to be consistent with the theoretical propositions, particularly the contingency hypothesis, the levels of statistical significance and explanatory power of the tests were marginal. The major limitations of the study , I feel, revolve around the conceptualization of the expectancy model. Three particular problems with the model used in this study are as follows.

First, it is unclear why the model provides for multiple extrinsic outcomes, but for only single intrinsic outcomes (for each of IV_a and IV_b). The issue is what "weights" should apply to each of the three classes of valence (IV_a, IV_b, and EV) incorporated in the model? Second, it is unclear whether the two classes of intrinsic outcomes (IV_a and IV_b) are intended to be viewed as mutually exclusive. It seems more plausible (as assumed in this study) that a particular intrinsic outcome could arise from both goal-directed behavior and goal accomplishment, and be valued differently according to each source. Third, and perhaps most important, the model fails to allow for the possibility of extrinsic rewards associated with goal-directed behavior. Both superiors and peers are as likely to react to job-effort (goal-directed behavior) as they are to job performance (goal-accomplishment) on the part of an individual.

The importance of this last issue is in connection with a reconciliation

17

of the behavioral and agency theoretical perspectives on MBE. In the agency model, MBE involves a variance investigation strategy which will yield a further signal regarding the subordinate's (agent's) performance, a signal which might indicate, for example, that an unfavorable variance was due to a lack of subordinate effort as opposed to factors beyond the subordinate's control. So long as the subordinate perceives that the signal generated by the investigation is not independent of the effort he exerts, then the promise of an investigation is motivating. Holmstrom (1979) showed that this is true whether or not unfavorable variances receive more attention than favorable variances. Baiman and Demski (1980) extended this result and showed that low-tail investigation strategies (i.e. MBE) are particularly appropriate for risk-averse subordinates. The uncertain outcome of the investigation gives it the characteristics of a "lottery" (Baiman and Demski, pp. 192-193) which risk-averse subordinates find particularly distasteful. Hence, it follows that the use of MBE would be associated with greater effort on the part of such subordinates to perform at a level which will avert the unfavorable variance which triggers the investigation.[6]

In the context of the expectancy model, the inclusion of a term, $P_3 EV_b$, might capture the motivational effects of the uncertainty associated with this investigation. EV_b represents an extrinsic reward associated with goal-directed behavior, and P_3 is the probability that such a reward will be forthcoming, or, in the agency context, the probability that the superior's investigation will reveal substantial goal-directed effort on the part of the subordinate, the basis for administering EV_b. P_3 would therefore capture the probabilistic outcome of the variance investigation process and we might predict that it would be positively associated with MBE.

Interestingly, Staw (1977) advocates a version of the expectancy model which specifically incorporates a P_3EV_b term. Unfortunately, the data from my study do not permit measurement of it. It seems that such an expanded model might provide a fruitful basis for some future integration of the behavioral and economic perspectives on MBE, which, on the basis of the results presented in Table 5, appears to be in widespread use in budgetary control systems.

19

TABLE 1

Means and Standard Deviations of Intrinsic Valences, and

Assessment of Differences among the Valences (Two-tailed, paired sample t-test)

Outcome[+]	IV_b Mean	IV_b Std. Dev.	IV_a Mean	IV_a Std. Dev.	Differences Among Means t	Differences Among Means p	r*
1.	2.06	1.17	2.16	0.97	−1.00	n.s.	0.45
2	1.38	1.34	0.90	1.53	3.60	< 0.01	0.49
3	1.15	1.24	0.91	1.18	2.20	< 0.05	0.52
4	2.01	1.18	0.87	1.25	9.08	< 0.01	0.35
5	1.79	0.84	1.84	0.98	−0.58	n.s.	0.03
6	0.71	1.28	0.32	1.34	3.49	< 0.01	0.55
7	0.54	2.00	2.15	1.04	−10.70	< 0.01	0.56
8	−1.77	1.54	−1.07	1.54	−4.31	< 0.01	0.33

+ See appendix for description of outcomes numbered in table.

* All correlations significant at p < 0.01, except r   0.03, n.s.

TABLE 2

Means and Standard Deviations of Extrinsic Valences

Outcome*	Mean	Std. Dev.
1.	3.09	0.84
2.	0.70	0.88
3.	3.10	0.87
4.	2.25	1.10
5.	3.16	0.81
6.	2.30	0.83
7.	1.41	1.07
8.	−1.09	1.30
9.	1.79	1.27

* See appendix for list of outcomes numbered in table.

TABLE 3

Means and Standard Deviations of Probabilities/Instrumentalities

and Correlations among P_1 items

i	P_{2i}		P_1 (three items)	
	Mean	Std. Dev.	Mean	Std. Dev.
1.	0.66	0.24	0.82	0.13
2.	0.48	0.19	0.80	0.12
3.	0.64	0.22	0.72	0.18
4.	0.48	0.17		
5.	0.60	0.18	Correlations between	
6.	0.67	0.22	three measures of P_1	
			1	2
7.	0.45	0.21	2 0.41*	
8.	0.28	0.22	3 0.35*	0.34*
9.	0.55	0.19	* p < 0.01	

TABLE 4

Mean, Standard Deviation, and Range of Final Motivation Measure

Std. Dev.

3.10

TABLE 5

Descriptive Statistics for MBE

Item		Mean	Std. Dev.
— Unfavorable variances receive more managerial attention than favorable variances		0.81	0.14

	* Unfavorable variances	* Favorable variances	Difference	$\underline{t}^{+}$	$\underline{p}$
I am required to trace the cause of * variances to groups or individuals within my department.	0.58	0.34	0.24	12.22	<0
My explanation of * budget variances is included in perform-ance reports.	0.63	0.53	0.10	6.16	<0
My superiors discuss budget items with me when * variances occur.	0.77	0.49	0.28	13.92	<0
I am required to submit an explanation in writing about causes of large * budget variances.	0.56	0.50	0.06	2.77	<0

Descriptive Statistics for Final Measure of MBE

Mean	Std. Dev.	Minimum	Maximum
0.299	0.125	0.03	0.67

+ A paired sample t-test was used.

TABLE 6

Descriptive Statistics for Milani and Hofstede Measures

Milani	32.99	5.40	14	42
Hofstede	5.46	1.38	1	8

25

TABLE 7

Correlations between MBE and Expectancy

Model Components   N   122

1.	10	12	−09	05	−01
2.	31***	11	−13	−01	−05
3.	10	−04	−01	−02	−11
4.	−03	−09	10	01	N/A
5.	09	−02	−06	08	N/A
6.	−01	06	−14	04	N/A
7.	08	03	−02	−09	N/A
8.	00	−13	−06	−11	N/A
9.	N/A	N/A	−23***	−11	N/A

* $p < 0.05$, one tailed   N/A: not applicable

** $p < 0.025$ +   see appendix for list of

*** $p < 0.01$ outcomes numbered in this

table

Decimal point suppressed.

26

TABLE 8

Correlation between MBE and motivation

under high versus low participation

Milani

High Participation

Low Participation   −0.23 (n=56)   −0.24 (n=53)

Difference*   t=1.83, p < 0.05   t=2.01, p < 0.05

* Tests performed under the null hypothesis that ρ   the sample correlation

   in the high participation group.

Table 9

Results of Regression based on Milani measure

α	9.96	0.28	35.51	< 0.01
β_1	−2.88	2.27	−1.27	< 0.10
β_2	0.50	0.28	1.77	< 0.05
β_3	3.23	2.27	1.42	< 0.10

R^2 0.05; $F_{(3, 117)}$ 2.07; $p < 0.10$; $N = 121$

Table 10

Results of Regression based on Hofstede Measure

Coefficient	Value	Std. Error			
α	9.99	0.28	^{c}	35.22	< 0.01
β_1	-2.79	2.28		-1.22	n.s.
β_2	0.05	0.28		0.17	n.s.
β_3	3.70	2.28		1.62	< 0.10

$R^2 = 0.03$; $F_{(3, 116)}$ 1.27; n.s.; N   120

Appendix

List of Outcomes

Eight intrinsic and nine extrinsic outcomes were used in this study. To avoid cluttering the tables, the outcomes were numbered as follows:

Extrinsic	Intrinsic
1. Pay Raise	1. Personal growth and development
2. High Pay	2. Setting higher standards for yourself
3. Respect from boss	3. Giving help to others
4. Respect from other employees	4. Time at work passing fast
5. Receiving more compliments	5. Feelings of security
6. Greater chances for independent thought and action	6. Setting higher standards for others
7. Fewer chances to make friends	7. Feelings of accomplishment
8. Special reward or recognition	8. Being tired
9. Promotion	

Footnotes

1. These raw scores were reversed and rescaled by subtracting five from all scores. As a result of this procedure, "neutral" responses score zero, while responses in the "desirable" direction score positively (one to four) and responses in the undesirable direction score negatively (minus one to minus four).

2. The complement of the score on the single item can be viewed as the probability that both favorable and unfavorable variances receive equal attention, or that favorable variances receive more attention than unfavorable (i.e., "reverse" MBE). In this sense, only scores in excess of 0.5 indicate the presence of MBE as defined in this study. Therefore, it might be argued that 0.5 should be subtracted from all scores on the single measure before its addition to the sum of the four difference scores. Such a procedure would result in subtracting a constant (0.1) from the final scores for MBE and would, therefore, not affect the results in any way.

3. Of the ten intercorrelations among the five items used to construct the MBE measure, five were significant at $p < 0.01$, one at $p < 0.02$, and the remaining four were either insignificant or signficant at no better than $p < 0.10$. These last four intercorrelations all involved the first of the four difference measures (see Table 5). On the basis of this result, an alternative, four-item MBE measure was constructed (omitting the uncorrelated item). However, none of the test results reported in the results section of this paper is affected by the choice between the four- and five-item measures.

31

4. The sample sizes for testing H_2 were reduced to $N = 121$ for tests involving the Milani measure and to $N = 120$ for tests involving the Hofstede measure. This was due, in each case, to improper completion of the participation measure involved.

5. None of the independent variables in either model was significantly intercorrelated. In the Milani model, the biserial correlation between participation (X_2) and MBE (X_1) was 0.09, and between participation and the interaction term $(X_1 X_2)$, -0.01. The product-moment correlation between MBE and the interaction term was 0.08. In the Hofstede model, these three correlations were, respectively, -0.03, 0.00, and 0.10.

6. In other words, for the risk-averse subordinate, a low-tail investigation strategy represents a penalty which the subordinate is motivated to avoid. Baiman and Demski also show that, for risk-tolerant subordinates, the investigation strategy (lottery) should be used as a "reward." This implies a high-tail strategy which will motivate a risk-tolerant subordinate to perform at a level which induces the investigation, that is, at a level which will produce a favorable variance.

References

Argyris, C., The Impact of Budgets on People. New York: Controllership
   Foundation, 1952.

Atkinson, J. W., "Towards Experimental Analysis of Human Motivation in Terms
   of Motives, Expectations and Incentives." in J. W. Atkinson, Editor,
   Motives in Fantasy Action and Society. New York: Van Nostrum, 1958.

Baiman, S. and J. S. Demski, "Economically Optimal Performance Evaluation
   and Control Systems." Studies on Economic Consequences of Financial and
   Managerial Accounting: Effects on Corporate Incentives and Decisions,
   supplement to the Journal of Accounting Research (1980): 184-220.

Bandura, A., "Social Learning through Imitation," in M. R. Jones, Editor,
   Nebraska Symposium on Motivation. University of Nebraska Press, 1962.

_____, "Influence of Models' Reinforcement Contingencies on the
   Acquisition of Imitiative Responses." Journal of Personality and Social
   Psychology (June 1965): 589-595.

_____, Social Learning Theory. New York: General Learning Press,
   1972.

Birnberg, J. G. and R. Nath, "Implications of Behavioral Science for
   Managerial Accounting." The Accounting Review (July 1967): 468-479.

Bittel, L. R., Management by Exception: A Systematizing and Simplifying of
   the Managerial Job. New York: McGraw-Hill, 1964.

33

Brownell, P., "Participation in Budgeting, Locus of Control and Organizational Effectiveness." The Accounting Review (October 1981): 844-859.

_____, "The Role of Accounting Data in Performance Evaluation, Budgetary Participation and Organizational Effectiveness." Journal of Accounting Research (Spring 1982): 12-27.

Bruns, W. J. Jr. and J. H. Waterhouse, "Budgetary Control and Organization Structure." Journal of Accounting Research (Autumn 1975): 177-204.

Bryan, J. F. and E. A. Locke, "Goal Setting as a Means of Increasing Motivation." Journal of Applied Psychology (June 1967): 274-277.

Calder, B. J. and B. M. Staw, "Interaction of Intrinsic and Extrinsic Motivation: Some Methodological Notes." Journal of Personality and Social Psychology (January 1975): 76-80.

Cherrington, D. J. and J. O. Cherrington, "Appropriate Reinforcement Contingencies in the Budgeting Process." Empirical Research in Accounting: Selected Studies, supplement to the Journal of Accounting Research (1973): 225-253.

DeCoster, D. T. and J. P. Fertakis, "Budget-Induced Pressure and its Relationship to Supervisory Behavior." Journal of Accounting Research (Autumn 1968): 237-246.

Galbraith, J. R., and L. L. Cummings, "An Empirical Investigation of the Motivational Determinants of Task Performance: Interactive Effects Between Instrumentality-Valence and Motivation-Ability," Organizational Behavior and Human Performance (August 1967): 237-257.

Georgopoulos, B. S., G. M. Mahoney, and N. W. Jones, "A Path Goal Approach to Productivity," Journal of Applied Psychology (December 1957): 345-353.

Hofstede, G. H., The Game of Budget Control. Assen: Van Corcum, 1967.

Hopwood, A. G.. "An Empirical Study of the Role of Accounting Data in Performance,Evaluation." Empirical Research in Accounting: Selected Studies, supplement to the Journal of Accounting Research (1972): 156-182.

Hull, C. L., "A Functional Interpretation of the Conditioned Reflex." Psychological Review (January 1929): 498-511.

Itami, H., "Evaluation Measures and Goal Congruence Under Uncertainty," Journal of Accounting Research (Spring 1975): 73-96.

Kenis, I., "Effects of Budgetary Goal Characteristics on Managerial Attitudes and Performance." The Accounting Review (October 1979): 707-721.

Lawler, E. E. and J. L. Suttle, "Expectancy Theory and Job Behavior." Organizational Behavior and Human Performance (June 1973): 482-503.

Holmstrom, B., "Moral Hazard and Observability." Bell Journal of Economics (Spring 1979): 74-91

Merchant, K. A., "The Design of the Corporate Budgeting System: Influences on Managerial Behavior and Performance." The Accounting Review (October 1981): 813-829.

Milani, K. W., "The Relationship of Participation in Budget-Setting to Industrial Supervisor Performance and Attitudes: A Field Study." The Accounting Review (April 1975): 274-284.

Otley, D. T., "Budget Use and Managerial Performance." Journal of Accounting Research (Spring 1978): 122-149.

Porter, M. and Z. Zannetos, "Administrative Regulation Versus Market Regulation in the Diversified Company," Sloan School Working Paper 987-78, MIT (April 1978).

Ronen, J. and J. L. Livingstone, "An Expectancy Theory Approach to the Motivational Impact of Budgets," Accounting Review (October 1975): 671-685.

Rotter, J. B., Social Learning and Clinical Psychology. Englewood Cliffs: Prentice-Hall, 1954.

_____, "Generalized Expectancies for Internal versus External Control of Reinforcement." Psychological Monographs: General and Applied (1966, whole No. 609): 1-28.

Spence, K. W., "Theoretical Interpretations of Learning." in F. A. Moss,
Editor, Comparative Psychology. Englewood Cliffs: Prentice-Hall, 1942.

Staw, B. M., "Motivation in Organizations: Toward Synthesis and Redirection"
in B. M. Staw and G. R. Salancik (eds.) New Directions in Organizational
Behavior. Chicago: St. Clair Press, 1977.

Swieringa, R. J. and R. H. Moncur, Some Effects of Participative
Budgeting on Managerial Behavior. National Association of Accountants,
1975.

Tolman, E. C., Purposive Behavior in Animals and Men. New York: Appleton
Century, 1932.

Winkler, R. L. and W. L. Hays, Statistics: Probability Inference
and Decision. New York: Holt, Rinehart and Winston, 1975.

CPSIA information can be obtained
at www.ICGtesting.com
Printed in the USA
BVHW091932020119
536891BV00021B/2378/P